STUDY TIPS FROM A TO Z

TAMARA LOFTON

ISBN: 979-8-89031-534-2 (sc)
ISBN: 979-8-89031-535-9 (hc)
ISBN: 979-8-89031-536-6 (e)

Because of the dynamic nature of the Internet, any web addresses or links contained in this book may have changed since publication and may no longer be valid. The views expressed in this work are solely those of the author and do not necessarily reflect the views of the publisher, and the publisher hereby disclaims any responsibility for them.

One Galleria Blvd., Suite 1900, Metairie, LA 70001
(504) 702-6708

DEDICATION........

To ideas
with no targeted direction.

To abstract thinking
with no unified cohesiveness.

To questions
with no immediate answers.

To inquiries
with no meaningful responses.

To aspirations
with no clear plan for execution......

I dedicate this book to YOU......

Love YA!

✦INTRODUCTION✦

My initial focus at Memphis State University was to merely pass all my classes and to maintain a B average in Music ED. Yes, I did graduate with a BSE Degree in Education by achieving only the minimum requirements to complete the degree.

After graduation, my first teaching position was in High School. Reality kicked in, while noticing vast learning gaps in my students. The culprit, poor study techniques.

I was thrust into other career opportunities totally unrelated to music. After enrolling in Nail Tech School, I quickly noticed that science and electrology classes were required. These were two of my least favorite subjects of all time. Desperate times required desperate measures! The birthing of sound study techniques was developed, tested, and utilized, passing my State Board Exam with a 96.3 final grade.

There were other learning hurdles to conquer and overcome after enrolling in Herzing College of Technology. Once again, I embraced this arsenal of tested methods, achieving a 3.85 GPA in Business Administration. What could I have accomplished, if I had only applied myself the first time around at Memphis State?

My prayer for you, is that you will experience great success as you apply these learning tips. Push past your limited mindset. Go above and beyond your present endeavors to reach your educational goals. These study tips will work no matter what course of study you may choose.

APPLY THE ACRONYMS

apply a phrase or single words
to retain educational rules or learning sequences

Being born in New York, reared in Illinois, experienced college in Tennessee, and migrated to Michigan, Mississippi, Louisiana, and Georgia, I realized a similar method of learning common to all of these locations. After interacting with the locals of these regions, basic life skills; potty training, tying our shoes, brushing our teeth, and washing dishes, had been taught, the exact same way.

This is an example of systematic teaching; the same subject, taught by the same method, produces the same life skill. We continue to practice these same skills in spite of where we may have come from.

To learn a skill, divorced from the application of it is useless, and a waste of time. The benefit is in the application! Think about what you already know. Who taught you? What method did they use? Was it beneficial? How long did it take to master it? Why did it take you so long to grasp the concept? What helped you to remember the lessons learned? Are you still using a form of these lessons today?

THE METHOD?

+ THE SKILL TAUGHT?
+ THE APPLICATION OF IT?
+ WHAT DID YOU LEARN?
+ WHAT DO YOU NOW KNOW?
+ HOW ARE YOU APPLYING WHAT YOU LEARNED?

The secret is in the application of "ACRONYMS." Incorporate a witty phrase into your learning arsenal to charge your memory potential.

Have you ever heard of these phrases?

"Lefty-Loosy-Righty Tighty"

(To loosen or tighten a screw).

"Please excuse my dear aunt, Sally?"

(Order of operations from Algebra class.)

"All Cars Eat Gas."

(Names of space notes on the bass clef).

These are all examples of acronyms. They are small groups of words or phrases; tools that boost the retention of important educational facts. They connect the learner to the subject matter.

Learning a skill divorced from the application of it is useless and a waste of time. That's why acronyms benefit the student and reinforce the skill of application.

BE A SPONGE

devour, drink in, soak up, suck in

I can still suffer from dehydration if I never pick up a gallon of water, remove the top from the container within my reach, and drink!

Absorb information like water. Information is worthless until ingested into our minds. With care and intentional development, it becomes knowledge that when stored, will greatly benefit us later. Just like nothing quenches our thirst like water, information turned into knowledge quenches our thirst to know more so that we can produce more!

COURAGE

audacity, bravery, daring, determination, endurance, fearlessness, firmness, fortitude, gallantry, grit, spirit, spunk, tenacity, backbone, guts, nerve, adventure, lion heartedness

"Remove the word can't from your mind and your mouth!" Haven't you noticed that you devise ways to do whatever you really want to do? It takes audacity to try new ways of learning. And it takes tenacity to try yet again after you haven't had much success the first time around. Discover the path that allows you to understand, then chart the course toward your own personal achievement. Success is failure turned inside out. Courage will get you there!

Discover the path that allows you to understand, then chart the course toward your own personal achievement.

DO YOUR DILIGENCE

intensity, application, constancy, exertion, earnestness, keenness, vigor

Everyone has a fighter on the inside. Tap into the fighter in you and apply a keen sense of willpower to reach your goal. You must promote yourself. Don't expect others to push you toward success. Be your own cheerleader!

ENERGIZE

animate, electrify, empower, invigorate, motivate,
stimulate, fortify, vitalize, build-up

Put forth great effort. Get excited about learning and
fortify your challenge to win. Focus on the prize!

FAMILIAR

intimate, commonplace, ordinary,
routine, conventional, known

Become familiar with vs. memorizing data. When you see, hear, smell, and experience something frequently, you tend to connect with it.

+ Hearing your name
+ Smelling your favorite food
+ A glimpse of your favorite color

These triggers send signals to your brain causing you to respond automatically. The response can be positive or negative, depending on the experience attached to the memory of it. This is the beginning, of becoming familiar with an unfamiliar subject or a new course of study.

+ How does it smell? Fearful or intriguing?
+ How does it make you feel? Comfortable or insecure?
+ What solidifies your connection to it? An urgent need to draw closer or are you totally disconnected?
+ How will you create positive learning experiences?

✦ Will you create learning games as an advanced study strategy, or will you cram the night before a major test?

A list of memorized facts has no connection to the experience. To embrace the total experience of sight, smell, taste, and touch integrates you and the event. It makes learning distinctly functional, like an appendage that has a purpose. Fact retention will become easy because you are no longer disconnected from the learning source.

GOALS

ambition, objective, intent, mission, mark, target

- ✦ Make a list of your goals.
- ✦ Improve your test-taking skills.
- ✦ Focus on reading comprehension.
- ✦ Enhance the ability to recall facts.
- ✦ Develop intentional focus and concentration.
- ✦ Probe subjects that you do not like.
- ✦ Devise a study schedule.

Set your mark, get ready, and go!

HAVE FUN

- ✦ Enjoy your learning experience!
- ✦ Create and play some learning games!
- ✦ Completely engage yourself in the process!

INSPIRATION

insight, motivation, stimulation

Who inspires you? Begin to study their journey
and implement some of their strategies to success.

IMAGINATION

Be creative and visualize yourself

in a better academic place.

JOURNEY

adventure, course, exploration, quiet,
route, travel, voyage

These things are progressive, but they do take time. Journey through the learning process. Don't jump through it. Enjoy the journey one step at a time!

KNOWLEDGE

ability, awareness, education, expertise, grasp, intelligence, proficiency, attainment, comprehension

Knowledge is power. How much power do you already possess? Create a learning file. Study areas that you are totally unfamiliar with. Place this information in a binder. The more knowledge attained; the more power possessed!

LEARNING STYLES

- ✦ Auditory-oral noise; Hear the lesson.
- ✦ Visual-notice everything; easily distracted; See the lesson.
- ✦ Kinesthetic-touch, hands-on the project; Feel the lesson.

For every learning style, there is a successful teaching method or strategy. Find a good model for your learning activity.

MAKE A LIST

compose, generate, prepare,
assemble, put together

Successful people make lists. These lists track where they are going and where they have been. Structure and order integrated with thinking allows you to retain and process more data.

NURTURE

care, rearing, training

Mayo Angelou said,
"When you learn, teach." You can coach a lot of
people after you've mastered the skill yourself.

OPEN

accessible, clear, free, unblocked,
uncluttered, expansive, unobstructed

Clear your mind from anything that blocks or restricts instruction absorption. Learning is a choice! We are the only thing preventing success in this area. Avoid the cabbage syndrome; nothing goes in and nothing comes out. Open your own mind to new possibilities.

PURSUE THE PROCESS WITH PATIENCE

go after, seek, badger, persevere, persist, trace,
track, hunt down, poke around, run after,
scout out, search for, search out

Have you ever played hide and seek outside in a vast open field? There are several places to hide, but if you listen to the giggling and whispering of the participants, you will soon discover their hiding places are not as concealed as you thought.

That's how learning is. There are many hidden clues in our textbooks and assignments leading us to the solution. These hidden treasures lead us to the path of discovery.

While pursuing our educational goals, we find that the process is sometimes slow, but with patience and persistence, the goal is within our reach. We must continue, follow the signs, and not give up. Understanding is the key that unlocks the door to the future.

Pursue and Conquer!

QUESTIONS

inquiry, investigation, examination

Question everything and as often as possible!
There are no unacceptable questions.

When you begin to ask and inquire, your mind begins to explore, discover options, and produce brain activity. The more questions you ask, the more stimulation to your brain. Critical thinking and subject awareness are activated to the point of clear understanding, and mental absorption.

+ Questions produce the ability to reason.

+ Questions broaden your level of awareness.

+ Questions create dialogue; a two-way conversation that will connect the teacher to the student.

+ Questions help you to grasp concepts that connect the unknown to the known.

READ

gather, interpret, scan, view, apprehend, comprehend, discover, pore over, examine, review, survey, take account of

Read to develop a well-rounded vocabulary. Read anything that you can get your hands on; a newspaper, a magazine, a calendar, a city map, a sale paper. All of these are good resources. The statistics from these periodicals will assist you with comprehension, math skills, reasoning skills, problem-solving, and more.

STUDY

consideration, debate, examination, inquiry, inspection, investigation, research, review, survey, analysis, attention, concentration, contemplation, meditation, pondering, questioning, reasoning, reflection, scrutiny, thought

+ Study often and always.
+ It is more than a casual glance.
+ Dig under the surface to find the real treasure.

THINK

consider, judge, envision, foresee, gather,
imagine, regard, surmise, visualize, plan for

I will use the mind that God has given me, and not depend on others to think for me. Think your thoughts all the way through. Engage yourself in thinking exercises. Brainstorm with a friend or classmate. Solve what-if scenarios. Challenge your thinking by exercising your mind.

UNPLUG

disconnect

Take an hour per day to disconnect from your electronic devices. You'll be surprised at the resourcefulness of your brain while giving that gray matter between your ears a break.

Your untapped thoughts will rise to new heights. Silence is golden and very therapeutic. Don't be afraid. Your devices will power up again as usual. And your ability to respond to every text that you missed during your quiet sabbatical will proceed normally.

VOCABULARY VICTORY

jargon, terminology, vernacular, language

Words greatly assist us in communicating with others. The broader the vocabulary, the more extensive the communication skills. Every subject has a language. When you master the language of any subject you will crack the code to understand.

I challenge you to learn just one new word every week. Define it, find the nature of its origin, spell it, pronounce it, and use it in a sentence. At the end of the year, you will have 52 new words as a point of reference. Then increase your word search to two words per week. The possibilities are endless. "Vocabulary Victory" is the language that liberates!

WORK

effort, endeavor, performance, production, struggle, assignment, drudgery, exertion, grind, muscle, push, striving, toil, travail, undertaking, elbow grease

To master any skill takes work, including "HOMEWORK". Use a little elbow grease and get those wrists moving as soon as you get home from school. You will find that the subject is fresh in your mind and that you will complete the assignment much quicker.

Don't waste time waiting until later. If you wait too late, you may fall asleep and then realize that you've not completed everything that your teacher wanted you to cover.

Work is not a dirty word. Rather, this productive word benefits all who will engage to finish the task before them. When challenges arise, and difficult situations try to hinder you, don't quit just work through it!

EXTRA WORK! EXTRA TIME! EXTRA SACRIFICE!

additional, extraordinary, further, new, other,
supplemental, more, surplus, lagniappe

Start doing more than is required. Go that extra mile. I remember a former student who was in trouble with her gym teacher. This student would not dress out for PE class and her grade was in jeopardy.

The teacher presented an opportunity for an extra credit assignment; to draw the layout of a soccer field. She was resisting and complaining, but I encouraged her to go that extra mile. She color-coded each of the team positions on the playing field, creating a beautiful replica of the entire soccer field. Her coach was so impressed that he displayed the soccer chart in the main athletic office for all to see.

YEARN TO KNOW AND TO GROW

hunger, craving, thirst, long, want, be desirous of, passionate, set one's heart on

Crave learning as you would do your favorite food or video game. Choose a subject that is your least favorite and purpose to research it Find a magazine article or search this subject on the internet. Then write a few paragraphs about what you discovered and place the findings in a binder. Over time, you will find that you've created a reading file that you can refer to. This will refresh your memory when preparing for a test or a project assigned to you unexpectedly.

ZEALOUS

dedicated, earnest, fanatical, fervent, passionate, devoted, eager, keen, obsessed, spirited

Become obsessed, fanatical, and passionate about mastering your assignments. Make them a priority Sunday night thru Thursday night. Weekends can be used as PTO (personal time off).

ACRONYMS

1. Lefty Loosy -Righty Tighty-
   "How Bolts and screws turn – Youtube - (Kongable, 2021)
2. Please excuse my dear aunt Sally-
   The Singing History Teachers - YouTube
   Order of Operations-PEMDAS
3. All Cars Eat Gas - https://www.piano-keyboard-guide.com/bass-clef.html
4. **LEARNING STYLES** -Auditory/Visual/Kinesthetic/www.mentor.edu.au
   The 4 types of learning styles/Mentor Education
5. **NURTURE**-When you learn teach(Maya Angelo) https://twitter.com>DrMayaAngelo>status Nov 28, 2018/May 29, 2014

Dictionary.com | Mea (Dictionary.com, 2023)nings & Definitions of English Words

Dictionary.com
https://www.dictionary.com

The world's leading online **dictionary: English definitions**, synonyms, word origins, example sentences, word games, and more. A trusted authority for 25+

Thesaurus.com: Synonyms and Antonyms of Words

Thesaurus.com
https://www.thesaurus.com

Thesaurus.com is the world's largest and most trusted online **thesaurus** for 25+ years. Join millions of people and grow your mastery of the **English** language.

BIBLIOGRAPHY

Angelo, M. (2018, November 28). Retrieved from Twitter: https://twitter.com>DrMayaAngelo>status

Cazaubon, M. (2009). *Piano-Keyboard-Guide*. Retrieved from Piano-Keyboard-Guide.com: https://www.piano-keyboard-guide.com/bass-clef.html

Dictionary.com, L. (2023). Retrieved from Dictionary.com: https://www.dictionary.com/

Kongable, M. (2021, June 1). *How Bolts and Screws Turn*. Retrieved from Mr. Kongable: https://www.youtube.com/watch?v=yzE3JoyTq7Q

Marinelli, A. (2021, October 08). Retrieved from The 4 Types of Learning Styles: https://www.mentor.edu.au/student-life/articles/the-4-types-of-learning-styles

Teachers, T. S. (2014, December 5). *The Singing History Teachers*. Retrieved from Youtube: https://www.youtube.com/watch?v=LwpUMJCSzec

www.ingramcontent.com/pod-product-compliance
Lightning Source LLC
Chambersburg PA
CBHW051335160726
47995CB00004B/1091